CPSIA information can be obtained at www.ICGtesting.com
Printed in the USA
BVIW121022020919
557354BV00013B/109